Deep Drawing

Dr. Nadhim M. Faleh

Engineering Collage

University of Mustansiriah

2017

PREFACE

The pervasive presence of metal forming and deep drawing in all aspects of engineering works and analysis is one of the manifestations of the industrial revolution that has characterized the 20th century. Every aspect of engineering practice, and even of everyday life, has been affected in some way or another by mechanical and electrical devices and instruments. Covers and containers in the food industry are perhaps the most obvious manifestations of this presence. However, many other areas of mechanical engineering are also important to the practicing engineer, from mechanical and industrial engineering. Mechanical engineers today must be able to apply effectively engineering sciences within the manufacturing operations in which they work.

Engineering education and engineering professional practice have seen some rather profound changes in the past decade. The integration of design and economic concepts in all engineering academic disciplines and the emergence of deep drawing technique as a central element of many engineering products and processes have become a common theme since the conception of this book. The principal objective of the book is to present the principles of metal forming, and deep drawing to an audience composed of mechanical engineering majors, and ranging from sophomore students in their first required introductory mechanical engineering course, to seniors, to first-year graduate students enrolled in more specialized courses in manufacturing, and production.

Contents

Chapter 1

Introduction

Machining, casting, welding and metal forming are the major methods of manufacturing. Metal forming is used for achieving difficult and complex profile products and improving the strength of the material. During forming, small material is wasted compare to other manufacturing methods. Sheet metal forming is done by many ways such as blanking, bending, spinning and deep drawing. Those methods are widely used for producing various products in different places of industry.

1.1 Metal forming

The parts manufactured by sheet metal forming are widely used in automotive and aircraft industries. Sheet metal working and Bulk forming are the two main types of forming methods based on raw material used in the operation. Rolling, extrusion and forging in bulk forming can be done cold, warm and hot. While the other forming method is sheet metal forming. In deep drawing, slim sheets of metal are shaped by applying force through dies. Sheet metal forming is very important for industry, because about %50 of metals is produced in sheet metals (Fig. 1.1).

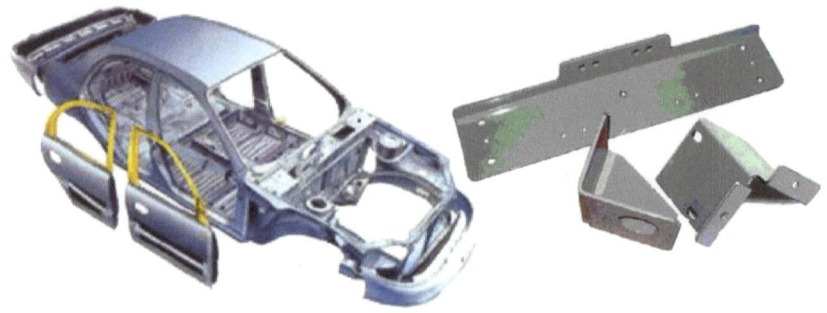

Fig. 1.1 Sheet metal forming.

1.2 Deep Drawing

Deep drawing operation is a sheet metal forming process in which a blank of sheet metal is radially drawn by the punch into a die under the mechanical action of a machine. It is thus a shape conversion process without metal losses. The operation is named "deep" drawing when the ratio of depth of the drawn part exceeds its diameter (Fig. 2.1).

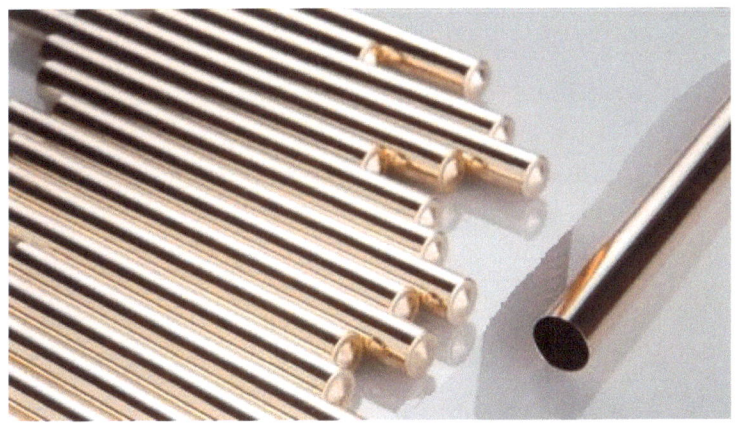

Fig. 1.2 Deep drawing product.

Deep drawing operation provides for us with a wide range of deep drawing capabilities for many metal products. Our wide variety of presses allows us to be responsive to our customers, based on product design and manufacturing solutions. The deep drawing process capability of normal case about 2.5 in depth-to-diameter ratio. There are a range of automated machines which can assist in keeping our product development depth to a maximum. By using new system, with in-house annealing capability we have total control of every aspect of the deep drawing process. There is no need to outsource the annealing process which would lengthen the operation time and drive up the cost of the product. The next figure shows many types of products are produced by deep drawing process. Many manufacturing industries have the potential to benefit from deep drawn metals for the manufacturing operation. This technique is often used for manufacturing small and large component parts. However, products of various shapes and sizes can be theoretically created through this the deep drawing technique. Everything from metal cans, kitchen sinks and cookware can be produced through this process. Sheet metal forming to make cup-shaped, box-shaped, or other complex-curved and hollow-shaped parts (Fig. 1.3).

Fig. 1.3 Deep drawing operation.

Deep drawing is one of the mainly important sheet metal forming technique. A flat plate is shaped into a 3-dimensions part by deep drawing. According to DIN 8584, the definition as follows: deep drawing is the tensile-compressive forming of a sheet blank to a hollow body open on one side or the forming of a pre-drawn hollow shape into another with a smaller cross-section without an intentional change in the sheet thickness.

1.3 Blank to Cup

The most popular metal forming methods available to manufacturers is deep drawing. Deep drawing starts with a flat plate of metal or disc. This operation is a sheet metal working process, in which products are manufactured from a sheet of metal called blank. This operation is called drawing, because the metal sheet is drawn into a die by a punch (Fig. 1.4).

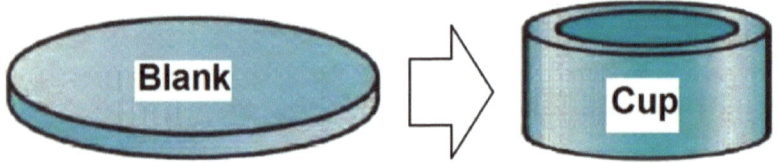

Fig. 1.4 The blank and cup.

1.4 Calculation of blank diameter

For final dimensions of drawn cup to be correct product;

(1) The blank diameter (Db) must be exact calculated (2)
Take blank volume = final product volume (3) To
facilitate your calculation, assume a negligible thinning of
part wall.

Method 1:

Volume of blank equal to the Volume of cup (Fig. 1.5):

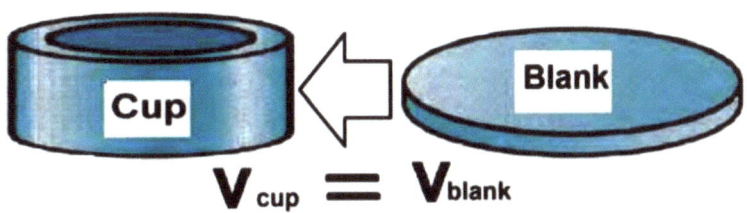

$$V_{cup} = V_{blank}$$

Fig. 1.5 The blank to cup.

If:

Blank Thickness (t) = Cup Thickness (t)

Then:

Blank Volume = Cup Volume

Blank Volume $=(\pi D_b^2 /4)(t)$

Cup Volume = ALL Volume + Inside Volume

Cup Volume $= (\pi D^2/4)\ h\ + (\pi d^2 /4)(h\text{-}t)$

See Fig. 1.6 and 1.7.

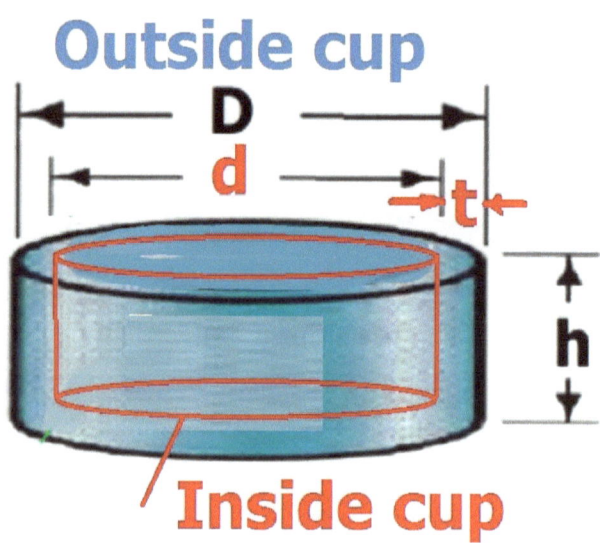

Fig. 1.6 The blank and cup.

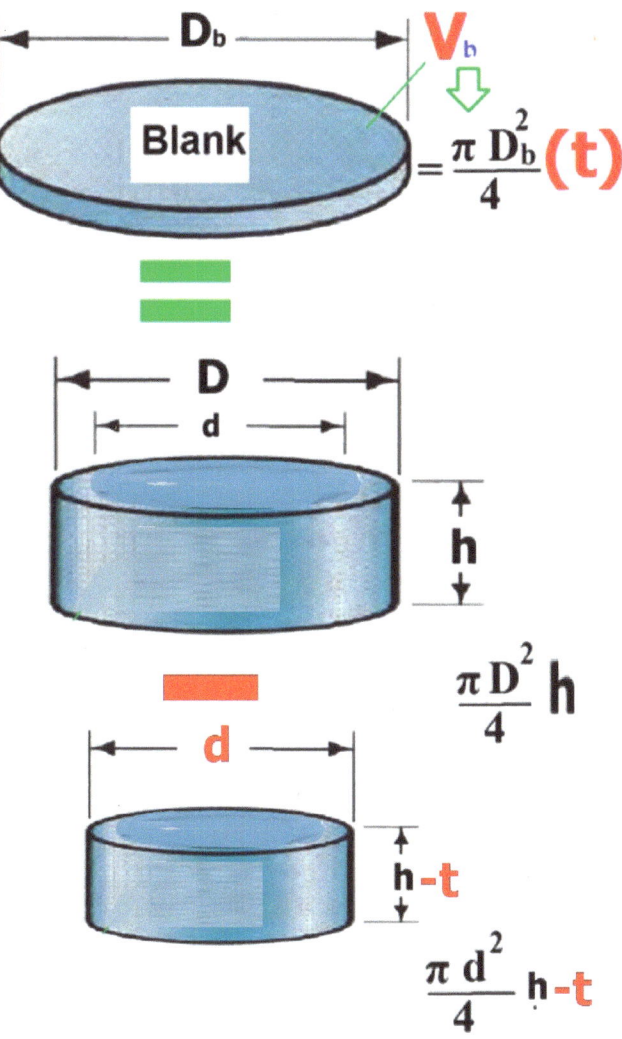

Fig. 1.7 The method1.

Calculate the blank diameter for the cup in Fig. 1.8, of

200mm and 10mm outside diameter and height

respectively. take sheet thickness (t)=2mm. Let; Blank

Thickness = Cup Thickness.

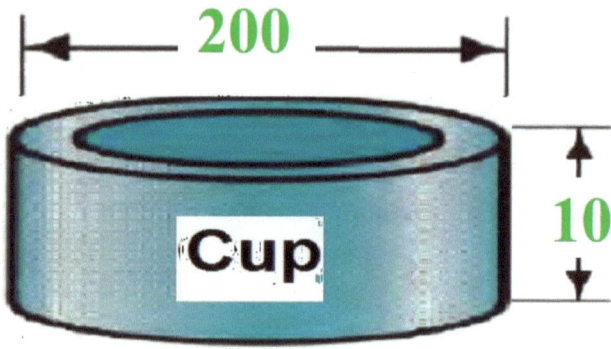

Fig. 1.8 The product.

Solution:

Blank Dia. = D_b

Cup Dia. = D = 200mm

Cup thickness (t)=2mm

Insid Dia = d =D- 2t = 196mm

Cup Height = h = 200mm

Cup Volume = $(\pi D^2/4)$ h $+ (\pi\, d^2\, /4)$ (h-t)

Cup Volume = $(\pi 200^2/4)$ 10 $+ (\pi\, 196^2\, /4)$ (10-2)

Cup Volume = 314000 +241252 = 72747 mm^3

Blank Volume $= (\pi\, D_b^2\, /4)$ (t)

$(\pi\, D_b^2\, /4)$ (t)= 72747 mm^3

$D_b^2 = 46336$ mm^2

$D_b = 215.3$ mm

Method 2:

Area of blank equal to the area of cup:

If:

Blank Thickness = Cup Thickness

Then:

Blank Surface Area = Cup Surface Area

Blank Surface Area $= \pi \, D_b{}^2 \,/4$

Cup Surface Area = case area + base area

Cup Surface Area = $\pi D \, h \; + \pi \, d^2 \,/4$

See Fig. 1.9.

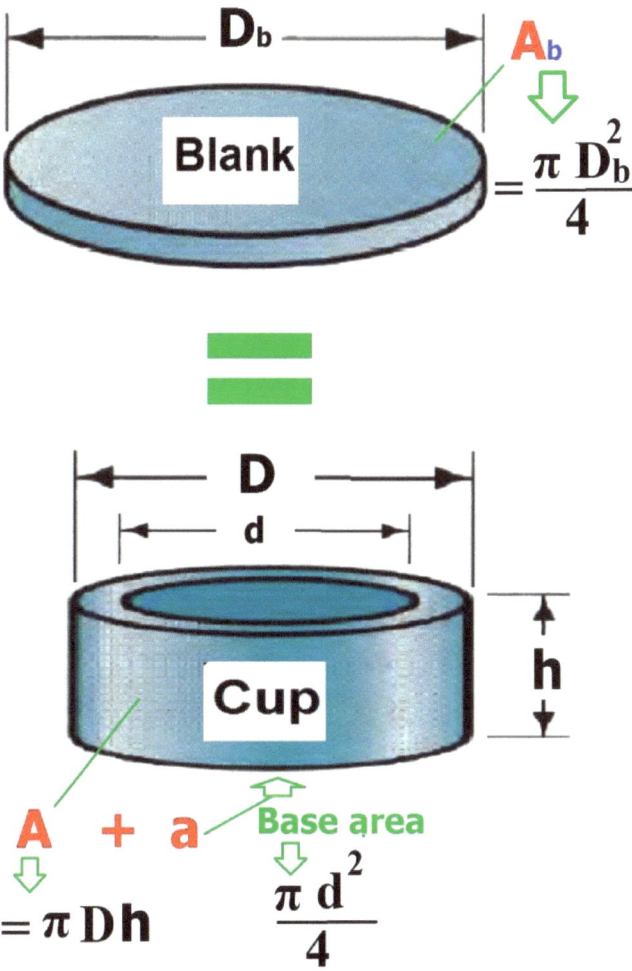

Fig. 1.9 The method2.

Calculate the blank diameter for the cup in Fig. 1.8 , take

sheet thickness=2mm. Let; Blank Thickness = Cup

Thickness.

Solution:

Blank Surface Area = Cup Surface Area

Blank Surface Area = $\pi \, (D)^2 \, /4$

$=0.78 \, D^2 \, mm^2$

Cup Surface Area = case area + base area

Cup Surface Area = $\pi D h + \pi \, d^2 \, /4$

Cup Surface Area = $\pi(200mm)(10mm) + \pi \, d^2 \, /4$

Inside diameter=Outside diameter $- 2 \times$ Thickness

d=D-2t

d=**200mm**-2(2mm)

d=196mm

Cup Surface Area = **6280mm^2** + **π (196mm)**^{2} /4

Cup Surface Area = **6280mm^2** + 30157mm2

Cup Surface Area = 36437mm2

Blank Surface Area = Cup Surface Area

0.78 D^2 = 36437mm2

D^2 = 46714mm^2

D= 216.1mm

See Fig. 1.10 and 1.11.

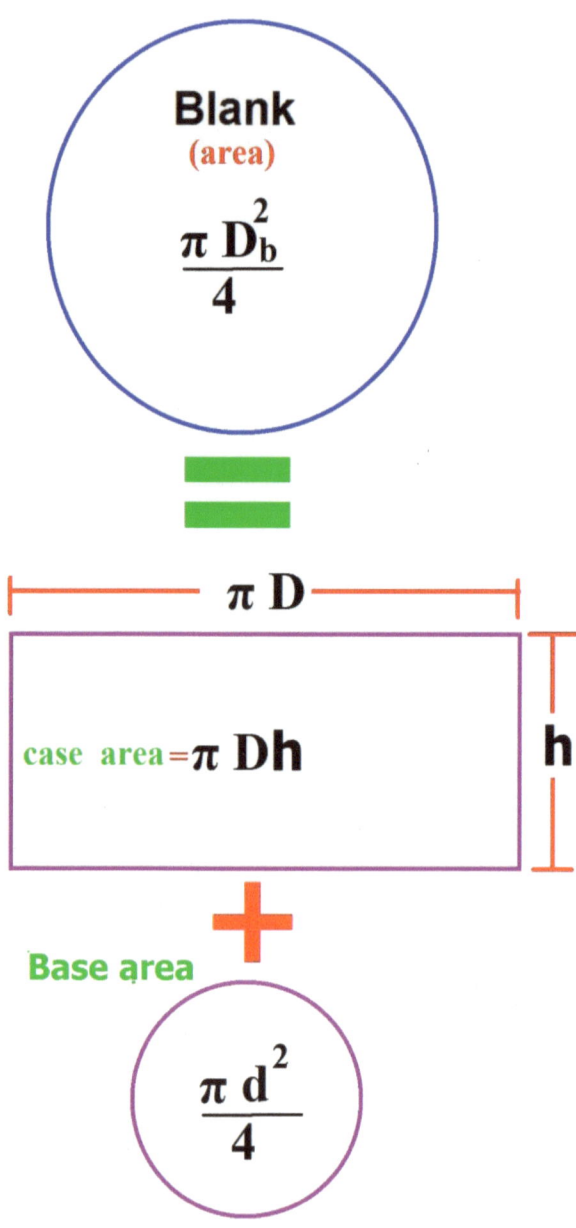

Fig. 1.10 The method of solution.

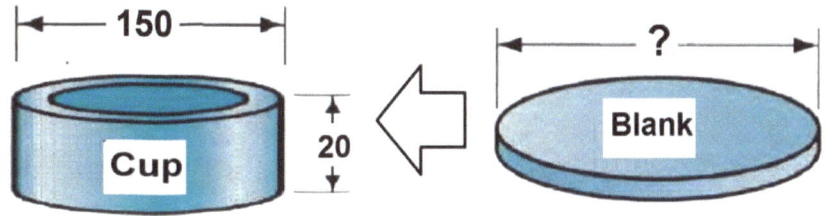

150

?

Cup

Blank

20

Fig. 1.11 The example.

Chapter 2
Introduction

In deep drawing working, there is no need for further forming as required for other operations. The time taken in deep drawing working is less than a half of that required in the other process. For carrying out deep drawing operation, the knowledge of geometry, dimensions, and properties of metal is most essential because nearly all shapes come from the development of the surfaces of a number of geometrical items such as cylinder and coin.

2.1 Punch and Die

Deep drawing using punch, die, holder and press machines to make a product of desired shape and size like a cup. Generally, metals used in deep drawing operation are galvanized iron, copper, brass, zinc and aluminum Fig 2.1.

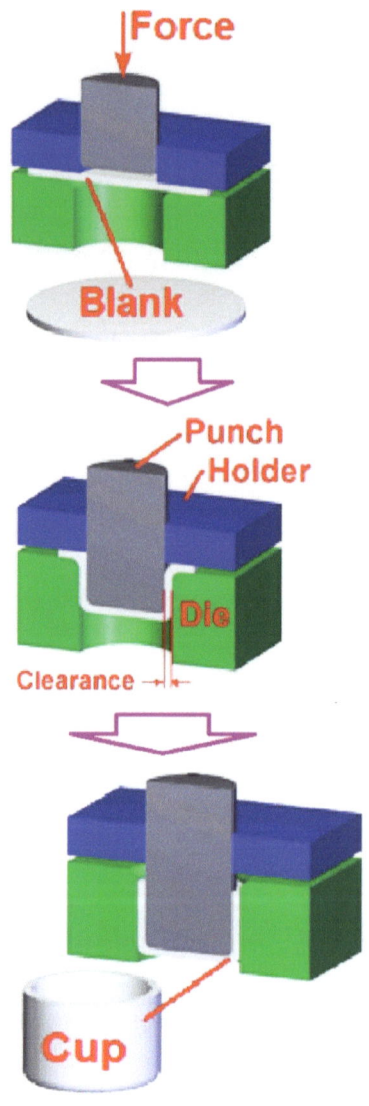

Fig. 2.1 The operation.

2.2 Punch and Die Calculation

Deep drawing of metal sheet is used to form cups and containers. A flat blank is formed into a cup by forcing a punch against the center of a blank. known as deep drawing to distinguish it from wire and bar drawing. Drawing of blank to produce a shaped cup is done by multi steps as follows: (1) punch near the end of stroke (2) punch contacts blank. (3) Starting blank and drawn part shown in lower views. The sheet metal which will be formed in this forming operation, has been set into die as seen in Figure 2.2 and apply force with a blank holder vertically. Then the sheet between blank holder and die is formed into cup under plastic deformation, by applying a vertical force by a punch which moves vertically to the blank.

2.3 Punch and Die Calculation

Clearance is the space between sides of punch and die in

deep drawing, a clearance (c) given by:

c = 1.1 t

where t = sheet metal thickness. Genially, clearance is

about 10% greater than sheet thickness. See Fig. 1.13

and 1.14.

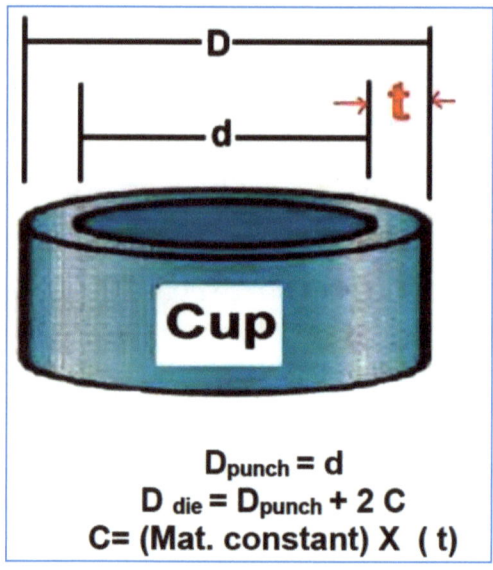

$D_{punch} = d$

$D_{die} = D_{punch} + 2\,C$

C= (Mat. constant) X (t)

Fig. 2. 2 The dimensions.

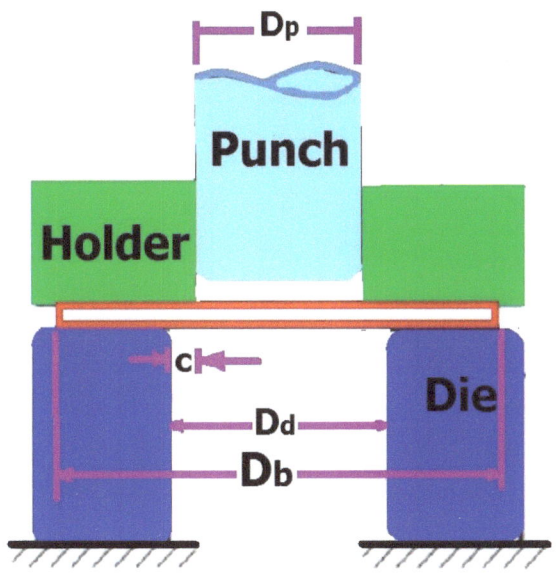

Fig. 2.3 The system.

QUIZ- A: Fig. (2) shows Cup produced by DEEP DRAWING operation. Calculate the following:

(1) Punch and Die diameter. Take thickness (t) =2mm and clearance= (1.1) x(t).

(2) The diameter of blank, assume thickness a constant during operation.

2.4 Drawing Ratio DR

$$DR = \frac{D_b}{D_p}$$

where Db = blank diameter;

Dp = punch diameter

- The limit: DR $\leq$ 2.0

2.5 Reduction r

- Defined for cylindrical shape:

$$r = \frac{D_b - D_p}{D_b}$$

Value of r must be less than 0.50

2.6 Drawing Force

Softer materials are much easier to drawn and therefore require less force to draw. Table 2.1 shows drawing force required for various materials and reductions in KN. The following is a table demonstrating the draw

force to percent reduction of commonly used materials.

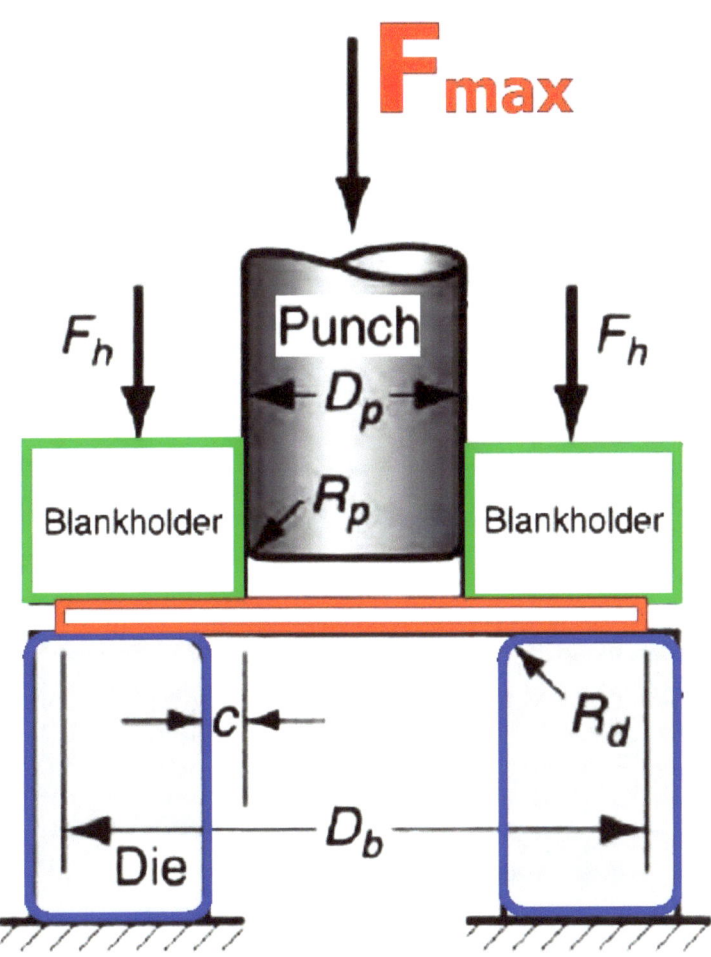

Fig. 2.4 Deep drawing system.

Table. 2.1

Drawing force required for various materials and reductions [kN]
[Rif: Todd, Allen & Alting 1994, p. 288.]

Material	Percent reduction			
	39%	43%	47%	50%
Aluminium	88	101	113	126
Brass	117	134	151	168
Cold-rolled steel	127	145	163	181
Stainless steel	166	190	214	238

2.7 Multi Stages

Typical products are in the food industry; packages of juice and drinks. Also in the automobile industry; doors and floor panels, etc. Final shape of a cup can be reached by multi drawing operations, these steps are called as redrawing. Multi drawing operations can be combined into one system using multiple die group in a hydraulic press. Deep drawing process is frequently used in the manufacture of cylindrical cups and similar objects in industry. Basically, the way is to draw cylindrical parts from circle disc. Special types of deep drawing system are liquid pressure forming and rubber die forming.

Blank changed into cup by deep drawing method is reshaped for transformation into another cup form with a narrower and deeper diameter (Fig. 1.4) by multi-stage of deep drawing operations. In the end of each step, hardening treatment occur.

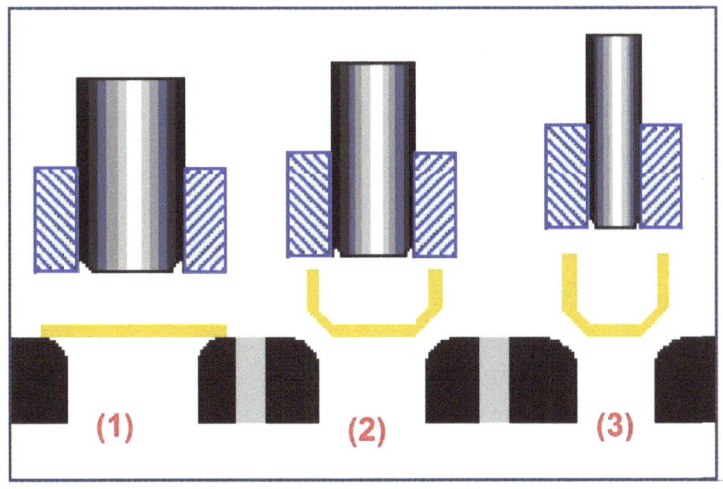

Fig. 2.5 Multi-stage operation.

2.8 Tool materials

Punches and dies are naturally made of **tool steel**, however **carbon steel** is cheaper, but not as hard and is therefore used in fewer severe applications, it is also common to see cemented carbides used where high wear and abrasive resistance is current. **Alloy steels** are usually used for the ejector system to kick the part out and in durable and heat resistant blank holders.

2.9 Lubrication and cooling

Lubricants are used to reduce friction between the working material and the punch on the one hand, and between the working material and die on the other. They also aid in releasing the part from the punch. Some types of lubricants used in this operations are heavy-duty emulsions, phosphates, **white lead**, and **wax** films. Plastic films casing both sides of the part while used with a lubricant will leave the part with an acceptable surface.

References

NARAYANAN, S.; KUMAR, K. GOKUL; REDDY, K. JANARDHAN; KUPPAN, P. (2006), CAD/CAM ROBOTICS AND FACTORIES OF THE FUTURE: 22ND INTERNATIONAL ONFERENCE, ALPHA SCIENCE INTERNATIONAL LTD., ISBN 81-7319-792-X

NADHIM M. FALEH, IRAQI JOURNAL OF CHEMICAL AND PETROLEUM ENGINEERING, PUBLISHER: BAGHDAD UNIVERSITY. ISSN: 19974884 YEAR: 2013 VOLUME: 14 ISSUE: 3 PAGES: 33-47,

MOROVVATI, M.R.; MOLLAEI-DARIANI, B.; ASADIAN-ARDAKANI, M.H. (2010), "A THEORETICAL, NUMERICAL, AND EXPERIMENTAL INVESTIGATION OF PLASTIC WRINKLING OF CIRCULAR TWO-LAYER SHEET METAL IN THE DEEP DRAWING", JOURNAL OF MATERIALS PROCESSING TECHNOLOGY, 210 (13): 1738–1747, DOI:10.1016/J.JMATPROTEC.2010.06.004

NADHIM M. FALEH, JOURNAL: ENGINEERING & TECHNOLOGY JOURNAL, PUBLISHER: UNIVERSITY OF TECHNOLOGY, ISSN: 16816900 24120758 YEAR: 2013 VOLUME: 31 ISSUE: 11 PART (A) ENGINEERING PAGES: 2030-2038

SALA, GIUSEPPE (JUNE 2001), "A NUMERICAL AND EXPERIMENTAL APPROACH TO OPTIMISE SHEET STAMPING TECHNOLOGIES: PART II — ALUMINIUM ALLOYS RUBBER-FORMING", MATERIALS & DESIGN, 22 (4): 299–315, DOI:10.1016/S0261-3069(00)00088-1

NADHIM M. FALEH, JOURNAL: ENGINEERING & TECHNOLOGY JOURNAL, PUBLISHER: UNIVERSITY OF TECHNOLOGY, ISSN: 16816900 24120758 YEAR: 2013 VOLUME: 31 ISSUE: 12 PART (A) ENGINEERING PAGES: 2242-2250

WICK, CHARLES; VEILLEUX, R. (1984), TOOL AND MANUFACTURING ENGINEERS HANDBOOK: FORMING, 2, SME, ISBN 0-87263-135-4

NADHIM M. FALEH, FATIGUE LIFE MODIFICATION, SCHOLAR'S PRESS (2016-06-07), ISBN-13: 978-3-639-86437-3, ISBN-10: 3639864379, EAN: 9783639864373.

TODD, ROBERT; ALLEN, DELL K.; ALTING, LEO (1994), MANUFACTURING PROCESSES REFERENCE GUIDE, NEW YORK: INDUSTRIAL PRESS INC., ISBN 0-8311-3049-0

Totten, George E.; Funatani, Kiyoshi; Xie, Lin (2004), Handbook of Metallurgical Process Design, CRC Press, ISBN 0-8247-4106-4